Meats & Protein

HEALTHY ME

Published by Smart Apple Media
1980 Lookout Drive, North Mankato, Minnesota 56003

PHOTOGRAPHS BY The American Meat Institute, Richard Cummins, Tom Myers, The National Pork
Board, Tom Stack & Associates (Inga Spence, Tom Stack), Unicorn Stock Photos (Joel Dexter, Jeff
Greenberg, Jean Higgins, Tom McCarthy, Karen Holsinger Mullen, Jim Shippee, Aneal E. Vohra)
DESIGN BY Evansday Design

Library of Congress Cataloging-in-Publication Data
Kalz, Jill.
Meats and protein / by Jill Kalz.
p. cm. — (Healthy me)
Summary: Describes various foods in the meat and protein food group and their role in
human nutrition. Includes a recipe for peanut butter balls.
ISBN 1-58340-298-5
1. Meat—Juvenile literature. 2. Proteins in human nutrition—Juvenile literature.
[1. Meat. 2. Proteins. 3. Nutrition.] I. Title.

TX373.K35 2003
641.3'6—dc21   2002030902

First Edition

9 8 7 6 5 4 3 2 1

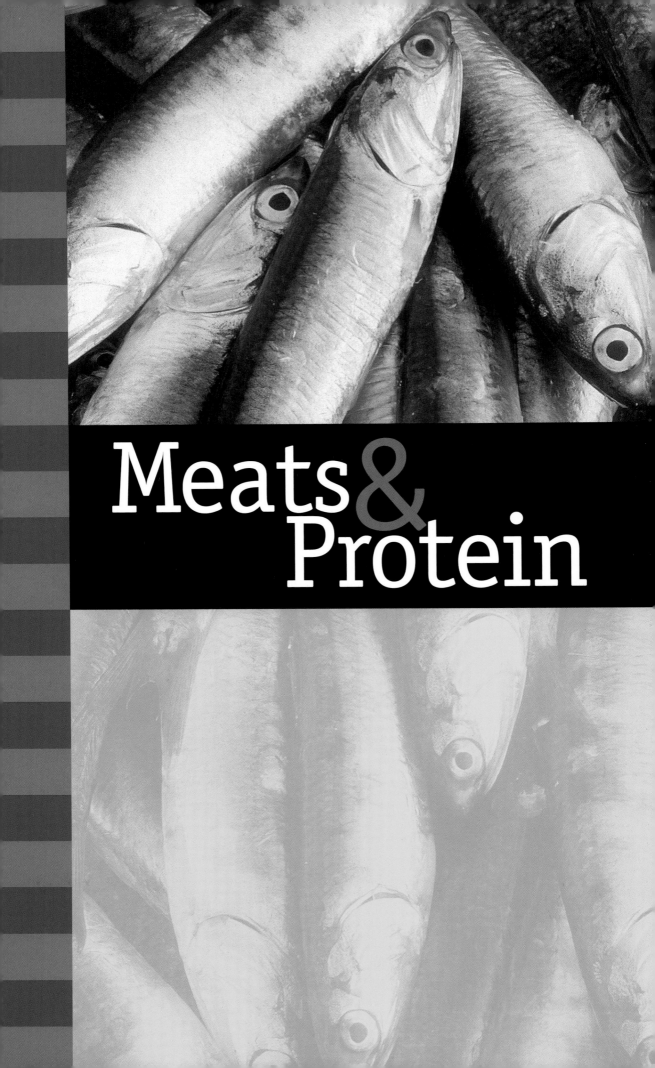

Meats&Protein

From the Farm

Some farmers grow plants. Others **raise** animals for their meat. Meat is animal **muscle**. Meat from cattle is called beef. Meat from pigs is called pork. Poultry is meat from birds such as chickens or turkeys.

Most of the meat you eat comes from huge farms. Cattle are raised on grassy fields. Pigs are raised in fenced yards called pens. Some chicken farms may have one million chickens! Farmers feed the animals a lot of food to fatten them up.

Turkey meat is called poultry. ⌃

Beef comes from full-grown cattle.

5

When the animals are the right age or size,

6 farmers take them to **processing plants**. There,

the meat is cleaned and checked for germs.

Then it is packed and sent to stores.

In some parts of the world, people eat meat from dogs, snakes, or alligators.

Many people buy meat in stores.

Meat-Eaters

All meat tastes different. And it comes in different colors. Beef is called red meat because it is red. Pork and poultry are white meats. When meat is cooked just right, it is moist and soft.

Butchers are people who cut meat. They cut beef into steaks. Hamburgers and hot dogs are also made from beef. Pork is cut into ham and bacon. Poultry is sold as whole birds or cut into parts.

A butcher cuts meat into pieces. ⌃

< Fresh, uncooked beef is red.

Some people hunt deer, rabbits, or ducks for their meat. This kind of meat is called wild game. Wild game has a dark color and a strong taste.

The average American eats 70 pounds (32 kg) of beef each year.

Iowa is home to more than 15 million pigs—but just 3 million people.

The Good Stuff

Meat has **nutrients** that everybody needs. The most important is protein. Protein helps your body grow. And it builds strong muscles.

Other foods have protein too. Fish, eggs, and nuts have a lot of protein. So do dried beans. Soybeans have three times more protein than beef. And most dried beans do not have a lot of fat.

Fish has nutrients your body needs. ⌃

< Protein builds muscles to help you run.

Some meats have a lot of fat. Too much fat can be bad for you. **Lean** cuts of beef and pork are good. Skinless chicken is even better.

14

People who do not eat meat are called vegetarians. They get protein from beans and nuts.

∨

Some hot dogs have a lot of fat.

Eating Right

All foods belong to one of five food groups.
Meats, fish, eggs, beans, and nuts belong to
the meats and protein group. Foods made from
milk belong to the dairy group. There are also
groups for fruits, vegetables, and grains.

Doctors say you should eat two or three helpings from the meats and protein group each day. A helping may be a hamburger. Two eggs. Or a handful of peanuts.

Peanuts belong to the meats and protein group. ^

< Meats come in many shapes, sizes, and colors.

It is important to eat foods from all of the food groups. Each group has nutrients your body needs. Meats help you grow. And they keep you strong!

Most sea fishermen catch fish by dragging
big bag-like nets behind their boats.

All of an egg's protein is in the clear,
watery part of the egg, called the white.

Peanut Butter Balls

**These peanut butter snacks have a lot of protein in them.
And they taste good, too.**

WHAT YOU NEED

One cup (275 ml) peanut butter
One-half cup (138 ml) honey
One teaspoon (5 ml) vanilla
Three cups (825 ml) puffed rice cereal
A mixing bowl and spoon
Waxed paper

WHAT YOU DO

1. Wash your hands.
2. Mix the peanut butter, honey, and vanilla together. Then stir in the cereal.
3. Use your hands to roll the mixture into small balls.
4. Put the balls on waxed paper. Chill in the refrigerator. Enjoy!

WORDS TO KNOW

lean has just a little fat or no fat at all

muscle a part of the body that helps animals and people move

nutrients things in food that keep your body healthy and growing

processing plants places where farm animals are killed and made ready to eat

raise to feed and take care of

Read More

Micucci, Charles. *The Life and Times of the Peanut*. New York: Houghton Mifflin Company, 2000.

Miller, Sara Swan. *Pigs*. New York: Children's Press, 2000.

Powell, Jillian. *Poultry*. Austin, Tex.: Raintree Steck-Vaughn, 1997.

Explore the Web

BURGERTOWN

http://burgertown.kidscom.com

PORK FOR KIDS

http://www.pork4kids.com

SKIPPY PEANUT BUTTER

http://www.peanutbutter.com/kidscorner_main.asp

Beans have more protein than many meats.